ROBOTICS AND ARTIFICIAL INTELLIGENCE

AN OVERVIEW OF ROBOT AND AUTOMATION TECHNOLOGY

FIRST EDITION

BY PRASUN BARUA

TABLE OF CONTENTS

INTRODUCTION

Welcome to the ROBOTICS AND ARTIFICIAL INTELLIGENCE! This book contains various types of topics on robotics and artificial intelligence. This is an overview of the robot and autonomous technology. Robot and autonomous technology is one of the rapid developing technologies contributing in autonomous industry significantly. By the virtue of these technologies, the autonomous industry and businesses become more efficient These technologies are contributing in various industries in terms of technology as well as economy. After reading this book, you will know about robot and autonomous technology.

This book covers topics such as robotics, artificial intelligence, importance of robotics in manufacturing, how robots are made, how artificial intelligence works, robotic arms, what is a PLC and how does it work. This the first edition of the book. It will be great pleasure if this book helps you to know about robot and autonomous technology. Thanks for reading the book.

CHAPTER-1: What is robotics?

Robotics is one of the branch of engineering which contains the conception, design, manufacture, and operation of robots. It connects with electronics, computer science, artificial intelligence, mechatronics, nanotechnology and bio engineering. Robotics are becoming increasingly predominant almost in every industry, from healthcare and manufacturing to defense and education.

A robot is a device which helps human to complete tasks. Robots are like tools. It is basically employed by the human hand through the coding it was programmed to follow. Robots can be used to supplement or replace human activity. It's an incredibly diverse field that produces all sorts of machines with numerous practical applications.

Features of Robotics

➢ Robots are constructed mechanically. It is designed with with different shape designed to complete a particular task.

➢ They can control and power the machinery by the help of electrical components.

➢ Robots can determine what, when and how to do something by the help of computer program.

Categories of Robotics

There are the three general categories of robotics:

Computation

Robots have a central processing unit called a controller which determines the actions they take in a given situation. These controllers can be programmed to finish tasks as simple as turning a screw or as complex as matching human social elegance and expressions.

Movement

Robots require specific mechanical parts to allow them to move freely without direct physical intervention from their human operators. Parts like wheels allow them to travel and motors which propel them. Other components like grippers allow them to interface with the world around them in a direct and targeted way.

Sensors

Robots can recognize their surroundings by the help of sensors. Sensors help robots to determine things like the size and shape of an object or detect heat, cold and other properties. These features of robots help the processors to collect data about the surrounding environment, then move accordingly.

Independent robots

Independent robots can follow their programming automatically without any direct physical intervention from a human operator. There are various practical applications of independent robots in society. They can execute any specific task without help of humans entirely. For example, we can say about the automation in factories. Robots can automatically accomplish tasks in factories without help of human.

Dependent robots

Dependent robots are non-autonomous robots which can interface with humans in ways that enhance their already existing actions. This can commonly be found in medicine and the field of prosthetic where robots are programmed to act in tandem with the human body itsel

CHAPTER-2: What is artificial intelligence?

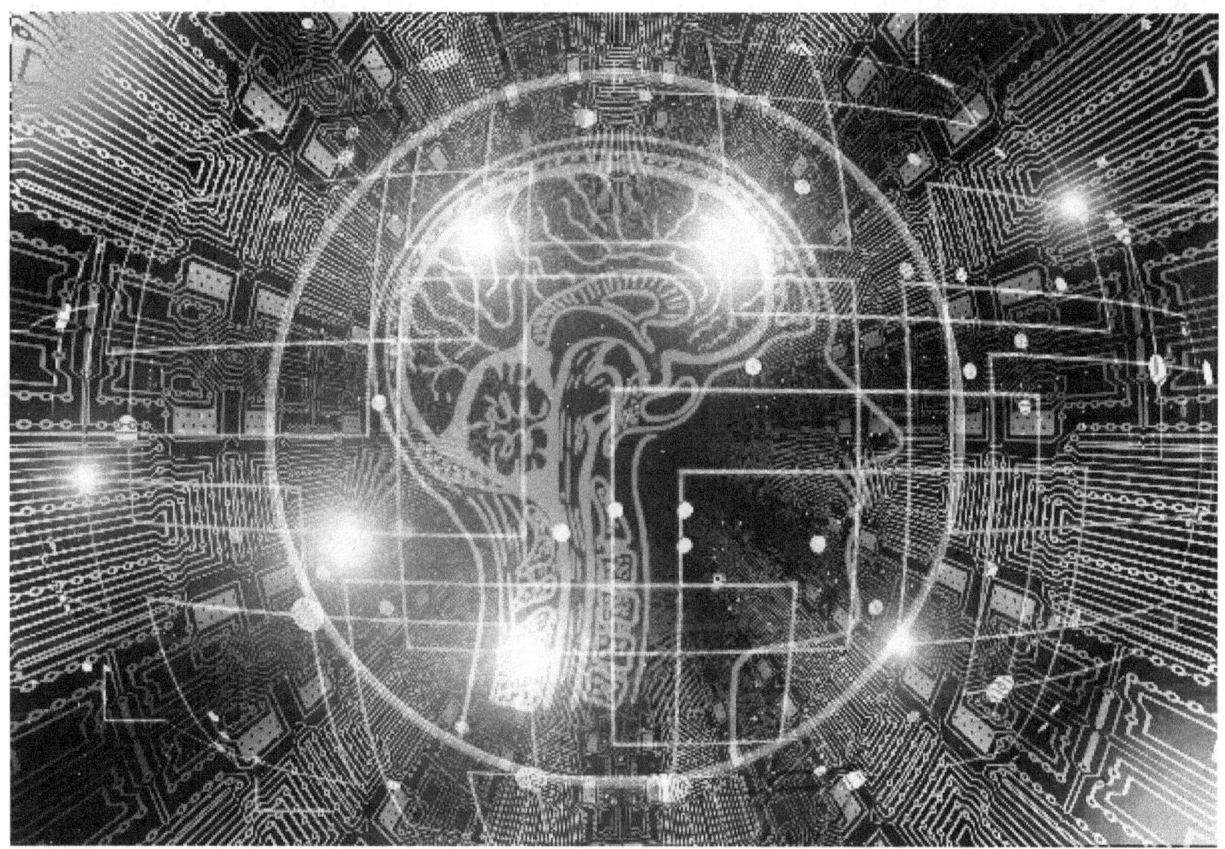

Artificial intelligence (AI) is the simulation of human intelligence in machines which are programmed to think like humans and mimic their actions. This is also applicable to any machine which can perform like a human mind such as learning and problem-solving. It is a wide branch of computer science which builds smart machines capable of performing tasks that typically require human intelligence. It is a various approached interdisciplinary science with advanced machine learning in every sector of the tech industry.

Artificial intelligence (AI) is the ability of a digital computer or computer-controlled robot to perform tasks related to intelligent beings. The term is frequently applied to the project of developing systems capable with the intellectual processes of human's characteristic like the ability to reason, discover meaning, generalize, or learn from past experience. Some programs have attained the performance levels of human experts and professionals in performing certain specific tasks, so that artificial intelligence in this limited sense is found in applications as diverse as medical diagnosis, computer search engines, and voice or handwriting recognition. The ideal characteristic of artificial intelligence is its ability to justify and take actions to achieve a specific goal.

Categorization of Artificial Intelligence

There are two different categories in artificial intelligence. They are weak and strong. Weak artificial intelligence system is designed to carry out a particular job. Weak AI systems include video games like the chess and personal assistants like Amazon's Alexa and Apple's Siri which can reply when we ask them any question.

Strong artificial intelligence systems can carry on the tasks like human. They are programmed in such a way that they can handle situations to solve problems without any human intervene. These types of systems can be found in applications like self-driving cars or in hospital operating rooms.

Applications of Artificial Intelligence

There are various applications of artificial intelligence. It is applied in many different sectors and industries. AI is being tested and used in the healthcare industry for dosing drugs and different treatment in patients, and for surgical procedures in the operating room. Other examples include computers which can play chess and self-driving cars. Each machine can evaluate any action they take, as each action will impact the end result. In chess, the end result is winning the game. For self-driving cars, the computer system can compute and analyze all external data for preventing a collision.

In the financial industry, artificial intelligence also plays a vital role. It can detect and flag activity in banking and finance like unusual debit card usage and large amount deposits which help a bank's fraud department. Artificial intelligence can also help modernize trading by making easier estimate of supply, demand, and pricing of securities.

CHAPTER-3: Importance of robotics in manufacturing

Nowadays, robotics plays a significant role in manufacturing industry. For getting maximum efficiency, safety and competitive advantage in the present market, automated manufacturing solutions are very helpful. Manufacturing robots automate repetitive tasks, minimize margins of error to negligible rates, and assist human workers to effort on more productive areas of the operation.

In manufacturing industry, robots perform various roles. For large scale repetitive processes — a robot can offer the speed, accuracy and durability. Robots are also used in manufacturing automation solutions to assist people with more complex tasks. The robot performs activities like lifting, holding and moving heavy pieces.

For remaining globally competitive and efficient, manufacturing companies use robotic process automation which offers an efficient, sustainable option to offshoring and accomplishing the skills gap in areas where it may be challenging to hire required employees. Manufacturing robots support employees to concentrate on innovation, modernization, efficiency and other, more complicated processes that eventually provides the foundation for growth and success. Enhanced productivity, upgraded worker safety and satisfaction, and a better bottom line can be obtained by using enthusiastic manufacturing automation solution.

Benefits of using robotics in manufacturing

➤ For creating efficiencies all the way from raw material handling to finished product packing, robots contribute effectively in manufacturing industry.

➤ Robotic equipment is highly flexible and can be customized to perform even complex functions.

➤ For operating 24/7 in lights-out situations for nonstop production, robots are programmed.

➤ For almost every size of company including small shops, Automation is comparatively cost-effective.

➢ For staying competitive in the market, manufacturers progressively require to hold automation as robotics are upgrading day by day.

➢ Robots can help to obtain ROI quickly, often within two years, counter weighing their upfront cost.

➢ Automation supports domestic companies to be price-competitive with offshore companies.

➢ As robots don't require climate control or lighting, they can help to save on utilities. They create cleaner spaces which contribute in the world where the importance of green manufacturing is increasing.

How robots help in manufacturing jobs

➢ By re-shoring more manufacturing task, robots help to create job in manufacturing industry.

➢ For maximizing workers' skills in other areas of the business, robots free up Robots help manufacturing companies to free up manpower.

➢ Robots can help to create jobs like engineering, programming, management and equipment maintenance. It protects workers from repetitive, routine and hazardous tasks.

➢ Due to decades of offshoring and robots remove the deficit, fewer skilled manufacturing workers are required in current labor market.

CHAPTER-4: How robots are made?

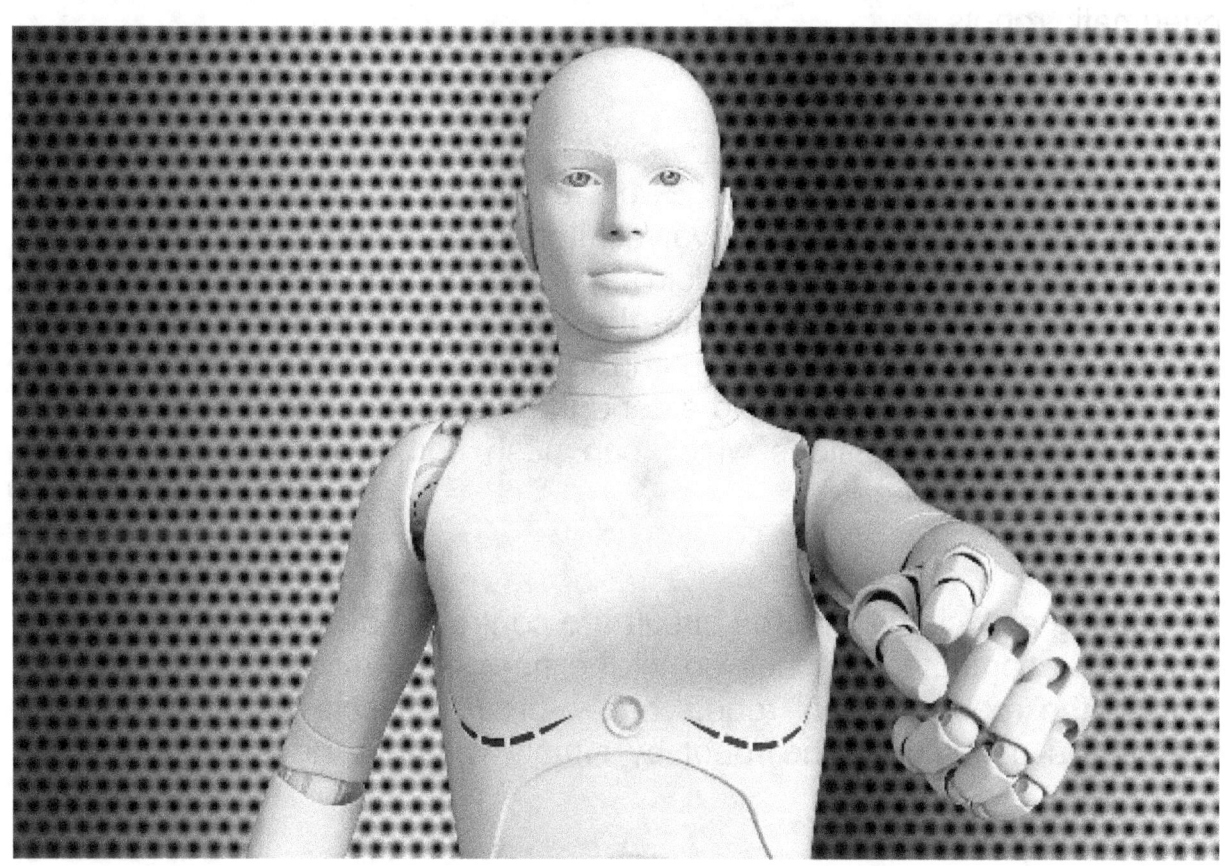

A robot basically contains a movable physical structure, a motor of some sort, a sensor system, a power supply and a computer "brain" which can control all of these elements. Robots are machines which can duplicate human and animal behavior. Most of robots have various qualities in general. Almost all robots have a movable body. Some only have motorized wheels and others have several movable segments, typically made of plastic or metal. Individual segments are connected together with joints like the bones of human body.

Robots spin wheels and pivot jointed segments with some sort of actuator. Robots use various types of actuator. Electric motors and solenoids are used in some robots as actuators. Hydraulic and pneumatic system are also used in different types of robots. Pneumatic is a system which is operated by compressed gases. In order to operate these actuators, a robot requires a power source. Most robots are operated either by battery or plug into the wall. For pressurizing the hydraulic fluid, a pump is required by a robot. An air compressor or compressed air tanks are required for pneumatic robots.

The actuators are electrically connected by wire into an electrical circuit. The circuit provides necessary powers to electrical motors and solenoids directly. By employing electrical valves, it triggers the hydraulic system. The valves regulate the pressurized fluid's path through the machine. In order to move a hydraulic leg, for example, the controller of the robot would open the valve leading from the fluid pump to a piston cylinder attached to that leg. The pressurized fluid would extend the piston, rotating the leg forward. Usually, robots use pistons which can push both ways for moving their segments in two directions.

Everything associated with the circuit are controlled by the computer of the robot. The computer switches on all the necessary motors and valves for moving the robot. Most of robots are re-programmable. You can change the robot's behavior of the robot by simply writing a new program to its computer.

Some robots have sensors which assist them to see, hear, taste and smell. The sense of movement is the most common robotic sense. A robot can monitor its own motion by utilizing this sensor. Slotted wheels are used in a standard design which are connected to the joints of the robot. An LED on one side of the wheel shines a beam of light through the slots to a light sensor on the other side of the wheel. The slotted wheel is turned on when the robot moves a particular joint. The slots break the light beam as the wheel spins. The light sensor reads the pattern of the flashing light and transmits the data to the computer. The computer can tell exactly how far the joint has rotated based on this pattern. These are the basic nuts and bolts of robotics. Scientists and engineers can create various types of robots by combining these elements in various ways.

CHAPTER-5: How artificial intelligence works?

Developing an Artificial Intelligence (AI) system is a cautious procedure of reverse-engineering human behaviors and competences in a machine and using its computational ability to exceed what human beings are capable of. Combining huge amounts of data with fast, iterative processing and intelligent algorithms, allowing the software to learn automatically from design and configurations in the data are the process to develop Artificial Intelligence. It works in following sub fields:

- In machine learning, it demonstrates a machine how to implicate and take decisions based on past experience. It classifies designs, analyses past data to conclude the meaning of these data points to reach a possible conclusion without involving human experience. This automation system helps in business by saving human time and assist human to make appropriate decision.

- In deep learning, it explains a machine to process inputs through layers in order to categorize, conclude and forecast the outcome.

- In neural networks, it processes data like human brains or neural cells. These are a series of algorithms which develops the relationship between various fundamental variables.

- In natural language processing, it helps a machine to understand what the user expects to communicate and it responds accordingly.

- In computer vision, it helps the machine to classify and learn from a set of images, to make a better output decision based on previous observations. Here, computer vision algorithms break down an image and read different parts of the objects. In this way, it tries to understand the image.

- In intellectual computing algorithms it simulates a human brain by analyzing text, speech, images and objects in a method that a human does and tries to give the desired output.

Artificial Intelligence can be made over a different set of components and will function as a consolidation of following fields:

- Philosophy

- ➢ Mathematics
- ➢ Economics
- ➢ Neuroscience
- ➢ Psychology
- ➢ Computer Engineering
- ➢ Control Theory and Cybernetics
- ➢ Linguistics

Philosophy

Philosophy assists machines to think and understand about the nature of knowledge itself. It helps to connect knowledge and action through goal-based analysis to achieve required outcomes.

Mathematics

Artificial Intelligence algorithms help to make accurate predictions of future outcomes for taking right decision. Here, the mathematical application, probability is used.

Economics

Economics explains how people make choices according to their desired results. It contains not only money, but also many important ideas like design theory, operations research and decision processes. By using mathematics, it demonstrates how these decisions are being made at large scales along with their collective results are. Intelligent Systems are developed by these types of decision-theoretic techniques.

Neuroscience

Neuroscience demonstrates how the human brain works. Artificial Intelligence tries to duplicate the same. Here, an obvious similarity is observed. Computers are millions of times faster than the human brain, but the human brain still has the advantage in terms of storage capability and interconnections. With advancement of computer hardware and more sophisticated software, it tries to achieve the intelligence level of human brain.

Psychology

Intellectual psychology views the brain as an information processing device and works based on principles and goals. This is as like as our own developed intelligence machine. Building algorithms by code helps to power the chat bots.

Computer Engineering

Computer engineering translates all concepts and theories into a machine readable language, computes it and give result which we can understand. Artificial Intelligence systems are developed with rapid advancement of the field of computer engineering. These are based on advanced operating systems, programming languages, information management systems, tools, and state-of-the-art hardware.

Control Theory and Cybernetics

For making a system intelligent appropriately, a system should be capable to control and modify its actions to produce the desired result. It is defined as an objective function. The system moves forward based on this. It can frequently modify its actions in various environment by using mathematical computations and logic to measure and improve its behaviors.

Linguistics

The formation of natural language processing is explained in Linguistics. It helps machines to understand our syntactic language, and give result so that everyone can understand it. For understanding a language, it is required to learn the structure of sentences and have a knowledge of the subject matter and circumstance.

CHAPTER-6: What are robotic arms?

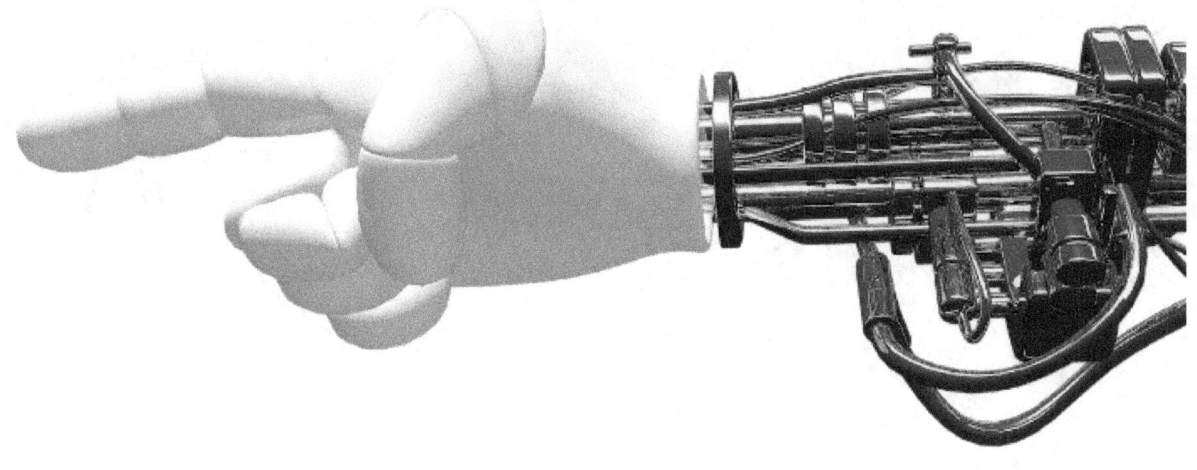

Like the human arms, robots also have arms. Robotic arms are mechanical arms which contain numerous parts for resembling the shoulder, an elbow and a wrist. These arms are programmed by machine language. For performing various types of functions like the human, machine language assists these robotic arms. These arms can execute a specific task or job quickly, efficiently, and extremely accurately. Robotics arms are basically motor-driven. They are used for the rapid, consistent performance of heavy and highly repetitive procedures over long periods of time. They are particularly valued in the industrial production, manufacturing, machining and assembly sectors.

A series of joints, articulations and manipulators are used in a typical industrial robot. In terms of mechanical perspective, it works together to closely resemble the motion and functionality of a human arm. A programmable robotic arm can be a complete machine in and of itself. It can function as an individual robot part of a larger and more complex piece of equipment.

In numerous industries and workplace applications, a good number of smaller robotic arms are used which are benchtop-mounted and controlled electronically. Some are floor-mounted and they are constructed from robust and long-lasting metal like steel or cast iron. It contains 4-6 articulating joints. The main joints on a robotic arm are designed like the main parts of the shoulder, elbow, forearm and wrist of a human body. It provides the mechanical structure and strength of the robot.

Industrial robotic arms can work at a specific speed and power. As these arms are preprogrammed by the machine language, they can perform the task within that specified speed and power. That's why these arms are programmed with a very safety-conscious. An end effector is a device located at the end of a robotic arm. It is designed to interact with the environment in which the robotic arm is to be used. This end effector is like the app which makes the robotic arm perform a variety of functions. The job of the robotic arm is to move the end effector from one location to the other based on the commands that the user sends via a control computer.

The design of the end effector depends on the applications for which the robotic arm is intended to be used. The end effector is also known as a robotic hand which can be designed to perform spinning, gripping, welding, and assembly operations. A robotic arm can be built-in with a variety of end effectors. One of the most commonly built-in end effectors is the one which resembles the human hand. It is used extensively for picking, gripping, and carrying different types of objects.

When robotic arms are deployed properly, they can increase production rates massively and perform the picking and placement tasks accurately. They can also perform lifting heavy-duty and relocation functions which is quite difficult even for groups of multiple human workers to carry out at any kind of pace.

With the technological advancement, the manufacturing costs of robotic components has decreased over the years, the past decade or so has seen a very rapid expansion in the availability and affordability of robots and robotic arms across a very extensive range of industries. Nowadays, they economically viable option for large-scale production lines delivering products with very high volumes of product. So, these robotic arms are significantly contributing in industrial production.

CHAPTER-7: What is a PLC and how does it work?

Full abbreviation of PLC is "Programmable Logic Controller". In order to operate under tough conditions of industries like dust, wet, dry and high temperature, a PLC is specially designed. It's a special type of computer. PLCs are broadly used in various types of industries because they're fast, easy to operate and are considered easy to program. Automatic industrial processes like mineral processing plant, wastewater treatment plant and manufacturing plant's assembly line use PLC machine for better productivity. Like a personal computer, a PLC also has a power supply, a CPU (Central Processing Unit), inputs and outputs (I/O), memory, and operating software.

There are numerous ways to program a PLC. It is programmed in from electromechanical relay based ladder logic and specially developed programming languages like BASIC and C. Nowadays, PLCs use one of the following 5 programming languages: Ladder Diagram, Structured Text, Function Block Diagram, Instruction List, or Sequential Function Charts. Users can view required data from manufacturing floor by using SCADA and HMI systems. These systems provide an interface for users to provide control input. PLCs are an important hardware component element in these systems. For communicating, monitoring and control automated processing such as assembly lines, machine functions and robotic devices, PLCs are used.

Following steps are followed in a PLC Scan Process:

> In the first step, cycling and monitoring of time are started by the operating system.

> Reading the data from the input module and checking the status of all the inputs. These processes are proceeded by CPU in this step.

➢ In this step, executing the user or application program written in relay-ladder logic or any other PLC-programming language are initiated by CPU.

➢ All the internal diagnosis and communication tasks are performed by CPU in this step.

➢ It writes the data into the output module according to the program results. In this way, all outputs are updated.

➢ As long as the PLC is in run mode, the process continues to run.